**Bibliographic information published by the German National Library:**

The German National Library lists this publication in the National Bibliography; detailed bibliographic data are available on the Internet at http://dnb.dnb.de .

**Imprint:**

Copyright © 2019 GRIN Verlag
Print and binding: Books on Demand GmbH, Norderstedt Germany
ISBN: 9783346032157

**This book at GRIN:**

https://www.grin.com/document/497172

Frank O. Ukoro

# Isolation of Bacteria and Fungi Associated with Palm Wine Sold in Gboko Metropolis

GRIN Verlag

**GRIN - Your knowledge has value**

Since its foundation in 1998, GRIN has specialized in publishing academic texts by students, college teachers and other academics as e-book and printed book. The website www.grin.com is an ideal platform for presenting term papers, final papers, scientific essays, dissertations and specialist books.

**Visit us on the internet:**

http://www.grin.com/

http://www.facebook.com/grincom

http://www.twitter.com/grin_com

# ISOLATION OF BACTERIA AND FUNGI ASSOCIATED WITH PALM WINE SOLD IN GBOKO METROPOLIS

## Ukoro, F. O.

*Department of Basic Sciences, Akperan Orshi College of Agriculture, Yandev, PMB 181, Gboko Benue State. NIGERIA.*

### ABSTRACT

Palm wine was assessed for microorganism of Bacteria and Fungi. The microorganism was assayed on *Lactobacillus, Bacillus, Staphylococcus aurus, E. coli, Streptococci, Pseudomonas aeroginosa,* with fungi; *Candida, Penicillilium, Yeast, Rhizopus, Mold, Yeast, Saccharomyces.* In Tarukpe joints only 11 microbial isolates were obtained, 6 Bacteria and 5 fungi with 42.9% and 33.3 % respectively. The rates of contamination and this may probably be due to poor handling and improper environmental sanitation around the joints. The occurrence of microbial isolate in palm wine supports the report made by Faparusi and Bassir (1971), Okafor (1987), Njoku *et al.,* (1991) which lend more weight to present findings. The presence of *Saccharomyces cerevesiae* collaborates with the earlier report of Bassir (1982).

**Keywords: Palm wine, Prevalence, Bacteria, Fungi.**

# Table of Contents

## INTRODUCTION

Palm wine is an important alcoholic beverage in West Africa, where it is consumed by more than 10 million people. Palm wine can be consumed in a variety of flavors varying from sweet unfermented to sour fermented and vinegary alcoholic drinks. There are many variations and names including "emu" and "ogogoro" in Nigeria and "nsafufuo" in Ghana. It is also called "bloody Mary" in Gboko when it is mixed with Stout beer or Vodkar (Saurabh, 2016).

The wine is produced from sugary palm saps. The most frequently tapped palms are raphia palms (*Raphia hookeri* or *R. vinifera*) and the oil palm (*Elaeis guineense*). Palm wine has been found to be nutritious (Iheonu, 2000).

Palm wine is characterize by it effervescence of gas resulting from sucrose content by the fermenting organism. Previous study on the microbiology of *E. guinensis* and *R. hookeri* have incremented several bacteria and yeast flora to be involved in fermentation process (Ogbuile et al. 2007), (Ezeronye and Okerentuba, 2007) reported the genetically and philosophical isolation of different yeast in palm wine.

In Nigeria (Owuana and Saunder, 1990) isolated both *S. cereviseae* and *Ampiculta specie* from palm wine whilst (Ezeronye and Okerntuba, 2000) reported the presence of *S. cerevisea*. (Enwefa *et al.,* 1992) however reported the presence of several general of yeast including, *Saccharomyces, Candidas, Endomycopsis, Hansenula, Kleodera, Pichai, Saccharomycoide* and *Schizosaccharomyces* in palm wine tapped from oil and raphia palms.

*Lactobaccilus plantariun* and *Leuconostoc mesentries* has been identify in the present work as dominant Lactic acid bacteria responsible for Sour taste of palm wine tapped from fell palm trees in Ghana (Uzochukwu *et al.,* 1984).

Palm wine is an adored drink of the palm belt dwellers (south Nigeria and south belt of Benue state). In Gboko town palm wine is sold in different locations. When this palm wine is brought for sale to the consumers, due to handling and poor hygiene different microorganisms are introduced into the wine. Sanitary deficiency may not be left out. Wine poured into an unsterile container (cup), uncovered and exposed into the air where flies, insect of sorts and vectors perch on can contaminate the wine and renders it unsafe for drinking. Microorganisms are ubiquitous. They can easily contaminate drinks sold to the consumers (Ogbulie, 1994). The aim of the present work was to isolate Bacteria and Fungi associated with palm wine sold in Gboko region of Benue state.

## Materials and Methods

### Study Population, Sample Collection and Techniques

The study was carried out in Gboko metropolis from palm wine sellers in different joints within the region. All the samples were collected at random and transported to the laboratory immediately and allow to settle. The total numbers of samples collected were twenty (20), five from each (joints) region around January, 2017. The study was conducted between January, February, 2017. The different locations of palm wine joints were sampled for microorganism. The samples were collected using randomized method, so that each sample of palm wine collected is given equal chance of been selected under the same conducive environment, maintaining same temperature of ice point ($0^0C$). Samples were

3

collected twice per week, five (5) each day, net collection per week was ten (10) samples. Five from Tarupke market, five from Abagu market, five from Ortese market and five from Gboko main market. From the above method twenty (20) samples were collected in two weeks. The wine was collected in a sterilize water bottle with ice inside it to maintain the normal temperature. The collection was done in Gboko metropolis from palm wine sellers. All samples were transported to the laboratory immediately for analysis.

## Media Preparation and Isolation of Microorganism

Three different agars, which includes potato dextrose agar (PDA), Manitol Salt agar and salmonella shigella agar were used for the analysis with four 250ml conical flask containing distill water twenty (20) Petri dishes, four (4) calibrated and five disposable Petri dishes were used as well. Sterilization was done by autoclaving at $121^0$C for 15 minutes. The plates were allowed to solidify before incubating at $37^0$C for 72 hours. Zero point one milliliter (0.1) each of the palm wine samples was introduce into a sterile Petri dishes with a pipette and the assay agar poured over it. Sterilization of agar was done at $121^0$C for 15minutes. The plates were allowed to solidify before incubating at $37^0$C for 72 hours. Pure cultures of the colonies were obtained by repeating streaking and maintain on agar slants which were kept at required temperature as stock culture. Inoccula were obtained from this slant for successive studies.

## BIOCHEMICAL CHARACTERIZATION
## Sugar Fermentation Test

The sugar solutions were 1% Glucose, Galatose, Lactose, Maltose, Fructose and Sucrose. The media used was 1% peptone water and test tubes containing distill water and a Durham tube for each of the test. The solution was sterilized at $121^0$C for 15 minutes. Phenol red was used as an indicator for acid production. Sterile wire loop was used to pick the colonies and dispensed into the test tube containing inverted Durham tubes to detect the present of gases by the yeast and incubated at $30^0$C for 4 days. With daily observation for color change by the indicator. Uninnoculated tubes serve as control (Gaffa and Azoro, 2005).

## Catalase Test

A wire loop was used to pick up the organism to be tested from a culture plate and placed in a drop of hydrogen peroxide in a clean glass slide.

Some organisms were seen to be positive, showing gas bubble production. The test was carried out to detect the present of Catalase which convert hydrogen peroxide to water and oxygen gas bubble. Negative reactions shows absent of gas bubble, positive reactions shows presence of gas bubble, a false positive reaction may be obtained if the culture media contains catalase, for instance if blood agar medium is used or if an iron wire loop is used. (Gaffa and Azoro, 2005).

## Coagulase Test

A sterile wire loop was used to pick a colony from an overnight culture and mixed with a normal saline place at the end of a clean glass slide. Drop of blood plasma was added and incubated at $37^0$ C for 1 – 6 hours. Clumping within 1 to 6 hours (1 – 6hrs) indicates a positive reaction (Ogbulie et al, 1994).

## Gram Staining

A thin smear was prepared and allows drying and heat fixed, the smear specimen was flooded with crystal violet and left for 30 seconds. This stains all the Bacteria violent. The smear was flooded with Lugols iodine and left for 30 seconds and washed with much water with a wash bottle. The slide was Flooded with acetone alcohol for decolorizing the gram negative organism within 3 seconds which might have been colored violent. Gram positive resisted the decolorizing agent with water. The slide was flooded with Safranin and left for 1 minute. The stain was Wash off with water and allow to dry. The slide was examined under the microscope at (x100) objective lens (Nezuami, 2014).

## Oxidase Test

Culture of the bacteria was made on an agar medium and allowed to grow. After growth a freshly prepared 1% tetra-methyl-p-phenylene-diamine dihydrochloride was poured on the plate so as to cover the surface. This is then decanted. Oxidase positive developed purple color rapidly while Oxidase negative do not develop the purple Color. (Gaffa and Azoro, 2005).

## Citrase Test

Cool medium was used with test organism from a saline suspension and incubated for 96 hour at $37^0$C. A positive result shows the present of turbidity which indicates growth, while negative results shows absent of turbidity, meaning there was no growth. (Gaffa and Azoro, 2005).

## Statistical Analysis

F- test will be use for this analysis.

## Result and Discussion

The result of the isolation of Bacteria and Fungi associated with palm wine sold in Gboko metropolis are presented in table 1-4.

| SITES | FUNGI ISOLATES | |
|---|---|---|
| 1. Tarukpe mkt | *Candida, Penicillilium, Yeast, Rhizopus, Mold* | |
| 2. Abagu (HQS mkt) | *Yeast, saccharomyces Mucor, Rhizopus, Mold,* | *Candida* |
| 3. Ortese mkt | *Saccharomyces* | |
| 4. Gboko mkt | *Yeast, saccharomyces* | |

**Table 1:** Shows the identification of Fungi associated with palm wine in Gboko metropolis, Tarupke and Abagu have six microbial isolates, Ortese has one while Gboko main market has two. Tarupke and Abagu abound with higher contamination of fungi due to poor sanitation in the areas followed by Gboko main market. Ortese has least contaminant due to high observants of hygiene level within the area. The presence of *Saccharomyces cerevesiae* collaborates with the earlier report of Bassir (1982).

**Table 2: Identification of Bacteria Associated with Palm Wine in Gboko Metropolis**

| SITES | BACTERIA ISOLATES | |
|---|---|---|
| 1. Tarokpe mkt | *Lactobacilus, Baccilus, Staphyloscoccus aurus,* *Streptococci, Pseudomonas aeroginosa,* | *E. coli, ,* |
| 2. Abagu (HQS mkt) | *E. coli, Streptococci, Staph aurus, Proteus sp.* | |
| 3. Ortese mkt | *Streptococci,* | |
| 4. Gboko mkt | *E. coli, Streptococci, Baccilus* | |

**Table 2:** shows the identification of Bacteria associated with palm wine in Gboko metropolis, Tarupke has seven microbial isolates and Abagu has four isolates, Ortese has one while Gboko main market has three. Tarupke and Abagu abound with higher contamination of Bacteria due to poor sanitation and handling in the areas followed by Gboko main market. Ortese has least contaminant due to high observant of hygiene level within the area.

**Table: 3 Comparison of the Rate of Microbial Prevalence in Gboko Metropolis**

| SITES | MICROORGANISMS | | | |
|---|---|---|---|---|
| BACTERIA ISOLATES | | (%) | FUNGI ISOLATES | (%) |
| 1. Tarokpe mkt | 6 | (42.9) | 5 | (35.7) |
| 2. Abagu (HQS mkt) | 4 | (28.6) | 6 | (42.9) |
| 3. Ortese mkt | 1 | (7.1) | 1 | (7.1) |
| 4. Gboko mkt | 3 | (21.4) | 2 | (14.3) |
| TOTAL | 14 | (100.0) | 14 | (100.0) |

**Table 3:** shows the comparison of the rate of microbial contamination among the four zones with palm wine in Gboko metropolis, Tarupke and Abagu have six microbial isolates, Ortese has one while Gboko main market has two. Tarupke and Abagu abound with higher contamination of fungi due to poor sanitation in the areas followed by Gboko main market. Ortese has least contaminant due to high observant of hygiene level within the area.

There is much difference between Tarokpe joints and Ortese. In Tarokpe the leading microrganism is Bacteria (*Lactobacilus, Baccilus, Staphyloscoccus aurus, E. coli, Streptococci, Pseudomonas aeroginosa*) with 42.9% and Fungi (*Candida, Penicillilium, Yeast, Rhizopus, Mold*) with 33.3%. While in Ortese we had 7.1% of streptococcus and 13.3% of yeast and Saccharomyces. The rate of prevalence was much in Tarokpe. It therefore means that Ortese has probably observed a reasonable level of hygiene in their environment. Unsterile water used during wine dilution may pose a threat to the consumer and same to public health. The hand of a food handler is link between the food and many other parts of his/her body. To prevent Bacteria from entering into eatable items hand should be washed properly before selling or serving the palm wine drink to its consumers. (Gaffa *et al.,* 2005).

**DISCUSSION**

The presence of some pathogens in palm wine may be as a result of transferred of these organisms by flies. Poor presentation and storage of this wine attracts flies that pitch on it and sometimes falls into the palm wine. This study reveals that palm wine samples collected in Gboko metropolis were contaminated with pathogens like *S. aureus, Salmonella* and *Shigella* spp. These pathogens are associated with some diseases like typhoid fever, urinary tract infection and food poisoning. Also, Tarukpe Market has the highest organisms (Rate of

7

contamination) with six genera and 16 replica. This may probably be due poor sanitation within the environment. The joint is operated under trees in a relatively scanty forest zones. Drinks are always open in gallons to prevent self autolysis and death of yeast. Adapting such practice allows or invites microorganism into the wine from around the environment. The occurrence of microbial isolate in palm wine supports the report made by Faparusi and Bassir (1971), Okafor (1987), Njoku *et al.,* (1991) which lend more weight to present findings. The presence of *Saccharomyces cerevesiae* collaborates with the earlier report of Bassir (1982). The identification of Bacteria isolate; like *Staphyloscoccus, Lactobacilus, Baccilus, Staphyloscoccus aurus, E. coli, , Streptococci, Pseudomonas aeroginosa,* and lots more from palm wine samples posses obvious public health question. The dominant of Bacteria is an evident of poor hygiene condition in some of the palm wine sample. These organisms may contaminate palm wine due to untreated water normally used in diluting the palm wine in other to increase the volume for profit maximization.

Base on the investigation in this study, identification of yeast species and other selective organism was based on biochemical and microscopic characterization.

The species of *Saccharomyces* are typical by their efficient capacity to convert sugar to ethanol. (Nwokeke 2001).

Furthermore, Saccharomyces cereviseae and other microorganisms such as *Lactobacilus, Baccilus, Staphyloscoccus aurus, E. coli, Streptococci, Pseudomonas aeroginosa* in carbohydrates utilization. Tables are presence for palm wine sold in Gboko metropolis.

However it was noted that palm wine during preparation for sale to consumers were contaminated due to the type of water and container used by the sellers to serve the existing market.

Table 3 shows that in Tarukpe joints only 11 microbial isolates were obtained, 6 Bacteria and 5 fungi with 42.9% and 33.3 % respectively. The rates of contamination and this may probably be due to poor handling and improper environmental sanitation around the joints. Microorganisms they say are ubiquitous they can be found everywhere even when glass cup use for serving customers are not sterile palm wine can still be contaminated. The table also shows that Tarukpe has the highest number of contamination and was far more than what was obtained in Abagu, Ortese and Gboko joints. Abagu became the next joint contaminated, reducing in that order. Ortese is the list contaminated which due to the fact that the wine is pure and not mixed called the up wine. The seller buys it in its natural form without mixing.

Therefore only 1 bacterium and two fungi were isolated (Streptococci and yeast and Saccharomyces).

There is much difference between Tarukpe joints and Ortese. In Tarukpe the leading microorganism is Bacteria (*Lactobacilus, Baccilus, Staphyloscoccus aurus, E. coli, Streptococci, Pseudomonas aeroginosa*) with 42.9% and Fungi (*Candida, Penicillilium, Yeast, Rhizopus, Mold*) with 33.3%. While in Ortese we had 7.1% of streptococcus and 13.3% of yeast and Saccharomyces. The rate of prevalence was much in Tarukpe. It therefore means that Ortese has probably observed a reasonable level of hygiene in their environment. Unsterile water used during wine dilution may pose a threat to the consumer and same to public health. The hand of a food handler is link between the food and many other parts of his/her body. To prevent Bacteria from entering into eatable items hand should be washed properly before selling or serving the palm wine drink to its consumers. (Gaffa *et al.*, 2005).

In Tarukpe the organisms includes; Candida, Saccharomyces species (cereviseae and besvengence) Yeast, Rhizopus and Mold. The colonial morphology of the organisms was studied. From Ortese joints only three organisms were isolated and they includes; Saccharomyces, Yeast, Streptococcus. The colore of the isolates in Ortese varies in color especially the fungi. Color ranges from cream to deep cream.

**CONCLUSION**

Palm wine is an adored drink of the palm belt dwellers sold in different joints around Gboko metropolis. There is danger in the patronizing palm wine generally. The danger may includes health challengers, the study attempt revealing the secrete of the implication of drinking in joints. The drink can go along way causing complication and contamination of the drink due to poor handling. The aim of the study was to identify microorganism of palm wine and to compare the rate of prevalence in different joints. The study has reviewed what poor handling of these drinks can cause and suggested a need for improvement to the consumers which pose threat to public health. It is essential to officially recognize the profession of wine tapping using community level or more formal institutions. Due to the important socio cultural role of palm wine, all interventions will benefit from close collaboration with traditional structures. Further research will help identify the factors associated with in-creased risk of injury and poor quality of palm wine.

## RECOMMENDATION

Organization of the palm wine tappers into groups can be the first step toward creating cooperatives to improve their market power and an ideal forum to increase awareness of their occupational risk. Such forums would also represent a great opportunity to provide alternative solutions (*e.g.*, modern climbing equipment). Although currently recognized by the Cameroonian government as "farmers," wine tappers represent a widespread industry which should receive more regulations to ensure the safety of both the tapper and the consumer in Nigeria. Such regulations would include

i. Quality control mechanisms in the palm wine production and provisions to ensure that safe climbing methods are used.

ii. In the future, tappers can be licensed (to tap wine) and their equipment inspected regularly.

iii. The health care system should develop surveillance and monitoring mechanisms for falls and associated injuries related to the palm wine industry.

iv. In Nigeria, formalized fermentation and industrial bottling of wine have already begun with promising results.

v. Finally, Nigeria should organized palm wine tappers associations (PTAs) to ensure that all tappers are registered with the association— an early form of regulation and official recognition of the industry.

**REFERENCE**

Ezeronye (2007). *A two-stage fermentation of cassava.* Nature (London). *183:* 620-621.

Enwefa, J. A (1992): *Preparation of Pito; a Nigerian fermented beverage. Journal Food Sci. Technol. 4:* 217 – 225.

Ezeronye T., Okerentuba R. (2000): *Benefits and uses of Microorganisms. In: M Julet (Ed) Introduction to Microbiology: A Case History Approach.* Third edition. Brooks/Cole – Thomson, Calif. USA. pp 270.

Gaffa T, Azoro C. (2005): *Bacteriology of Biology, catereers and food technology.* First published. Amana printing. Pp 95-96, 104-105.

Iheonu T.E (2000): effect of local preservative of plant origin of microbiology and shelf Life stability of palm wine. B. Sc thesis. Abia state University, Imo state university. Nigeria.

Owuana and Saunder (1990): *development of baking yeast.* World journal of microbiology and biotechnology, 10: Pp 199-202.

Okafor, N (1987*): Microbiology and Biochemistry of oil palm wine. Adv. Appl. Microbiol. 24:* 237 – 256.

Okerentuba, R. (2007): *Changes in the Physicochemical Characteristics of Processed and Stored Raphia hookeri Palm Sap (Shelf Life Studies). Am J Food Technol*; 2: 323- 326

Ogbulie, T. E., Ogbuilie J. N, and Njoku, H. O. (2007): comparative study on the shelf life statbility of palm wine from Elaeis guineensis and Raphia hookeri obtained from okigwe, Nigeria. African Journal of Biotechnology 6(7): 914-922.

Saurabh T. (2016). *Alcoholic beverage.* Academic press, New York. 33p.*website.*

Uzochukwu J. G. Wings, J; (1984): *Chemical and Microbiological studies on Congolese palmwine (Elaeis guineensis). East Afri. and Forestry J. 36:*311 – 314.

Faparusi, S I (1971): *A biochemical study of palm-wine from different varieties of Elaeis guineensis.* Ph.D. Thesis, University of Ibadan, Nigeria.

Nwokeke , N.V (2001): *palm wine preservation using traditional plant that have preservative bases.* West African Journal of Biotechnology. 5:1120118

Njoku, H. O, Ofuya, C. O, and Ogbulie, J. N. (1991): *production of Tempe from African yam bean* (Sphenostylis stenocarps Hams). Journal of food microbes 8:209-214.

Nezuami, N. (2014): elementary microbiology ofr community health and other allied health practitioners. First published O.J computer Pp. 14-15.

# YOUR KNOWLEDGE HAS VALUE

- We will publish your bachelor's and
  master's thesis, essays and papers

- Your own eBook and book -
  sold worldwide in all relevant shops

- Earn money with each sale

## Upload your text at www.GRIN.com
and publish for free